Cute Baby Animals
Dot-to-Dot Puzzles from 150-448 Dots

By Laura's Dot to Dot Therapy

How To Use This Book

Hi! We're so glad you're a lover of puzzles and dot connecting- we are too!

Connecting the dots in this book is simple- just relax and follow the numbers in consecutive order, drawing a straight line between each one. Dot 1 will connect to dot 2 and so on and so forth until there are no more dots to connect. There's always another dot and you'll always find it. Connect every dot to discover the beautiful images they create.

In case you get lost or can't find a dot, never stress- there's an answer key at the back of the book that will show you exactly where each dot connects to the next. If you want to color your images, we encourage you to do so! Feel free to try all different colors and coloring mediums for your images!

If you find any errors or omissions in this book, email us at Laurasdottodot@gmail.com and please let us know! We want you to have the best dot to dot experience!

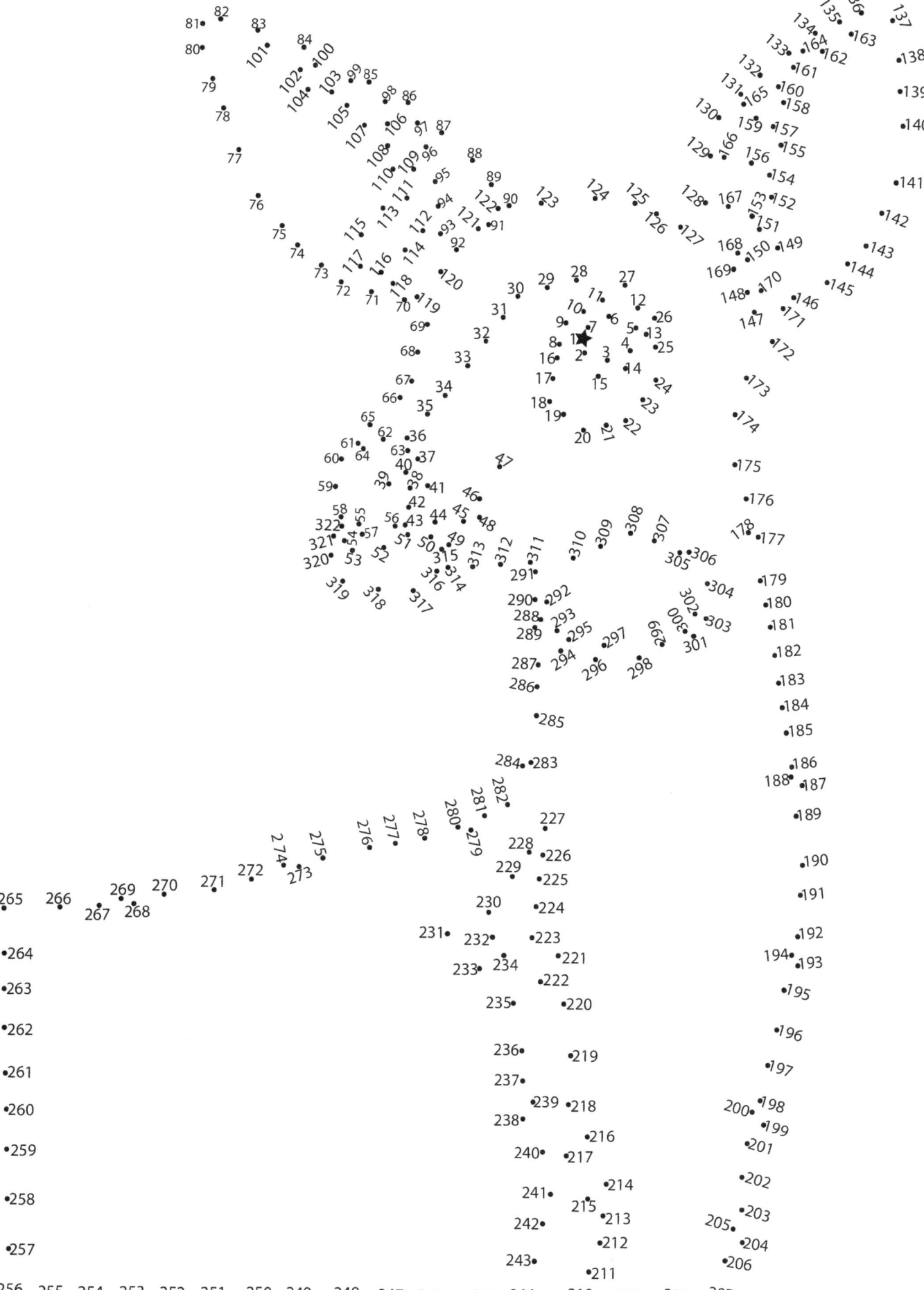

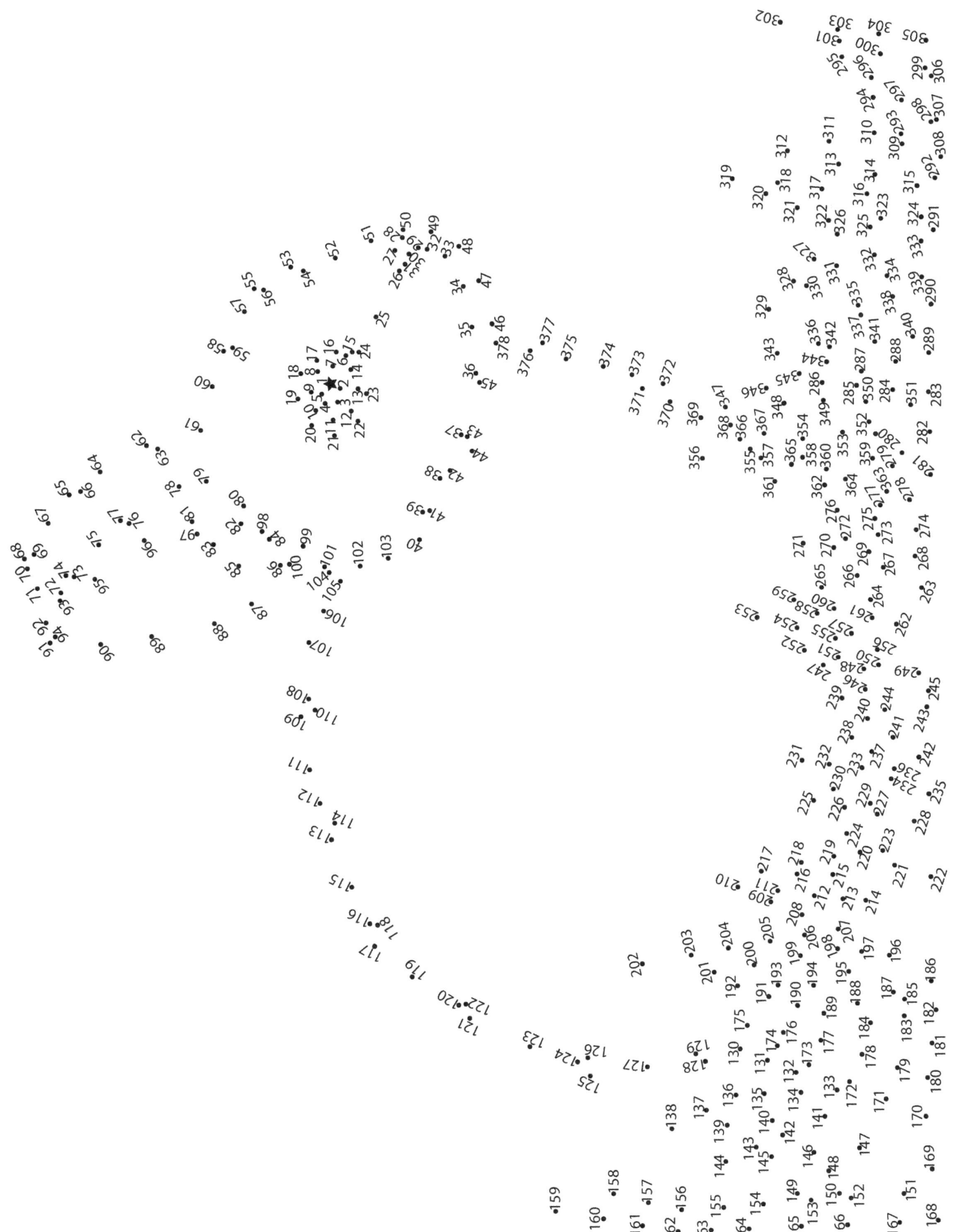

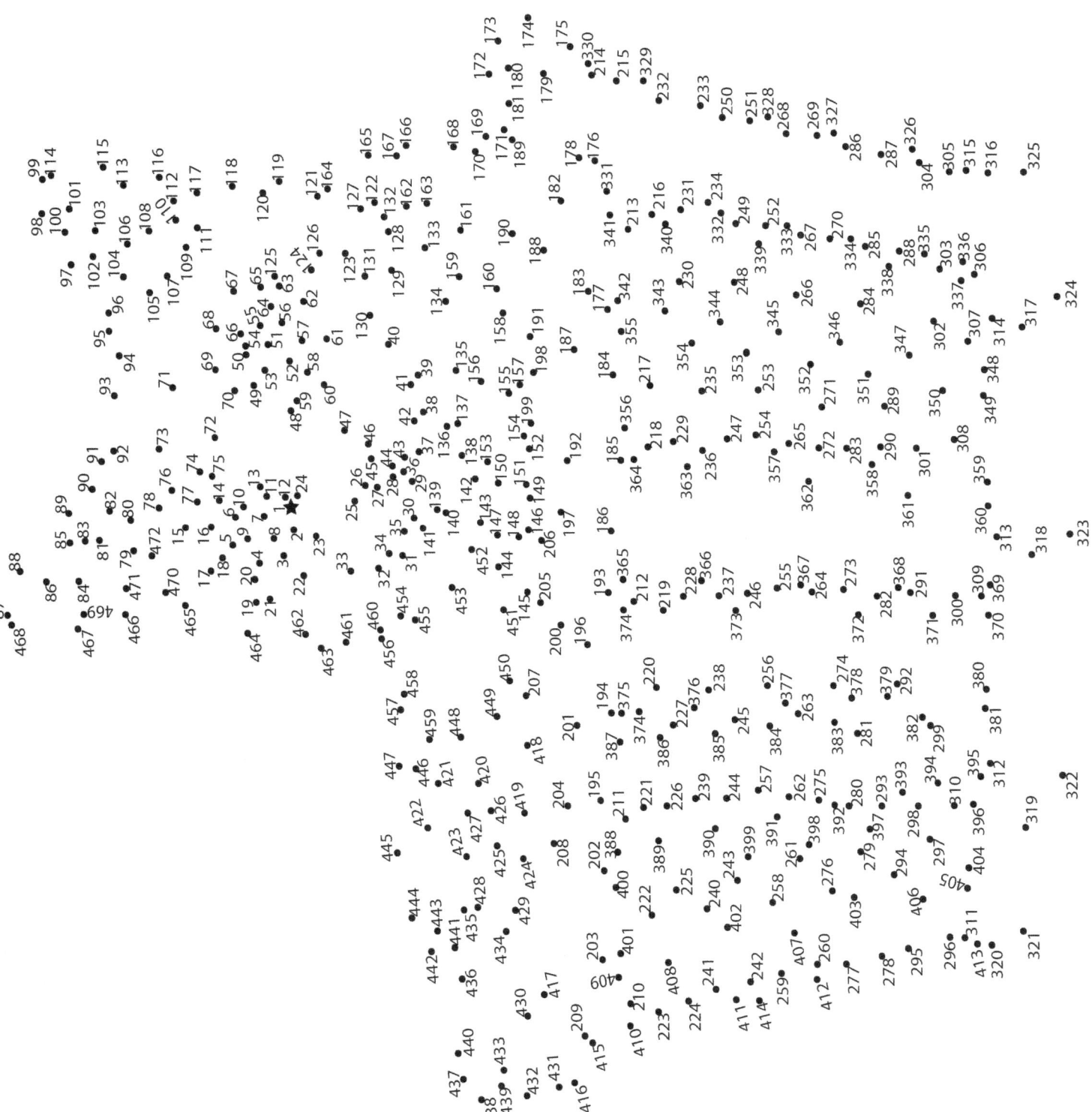

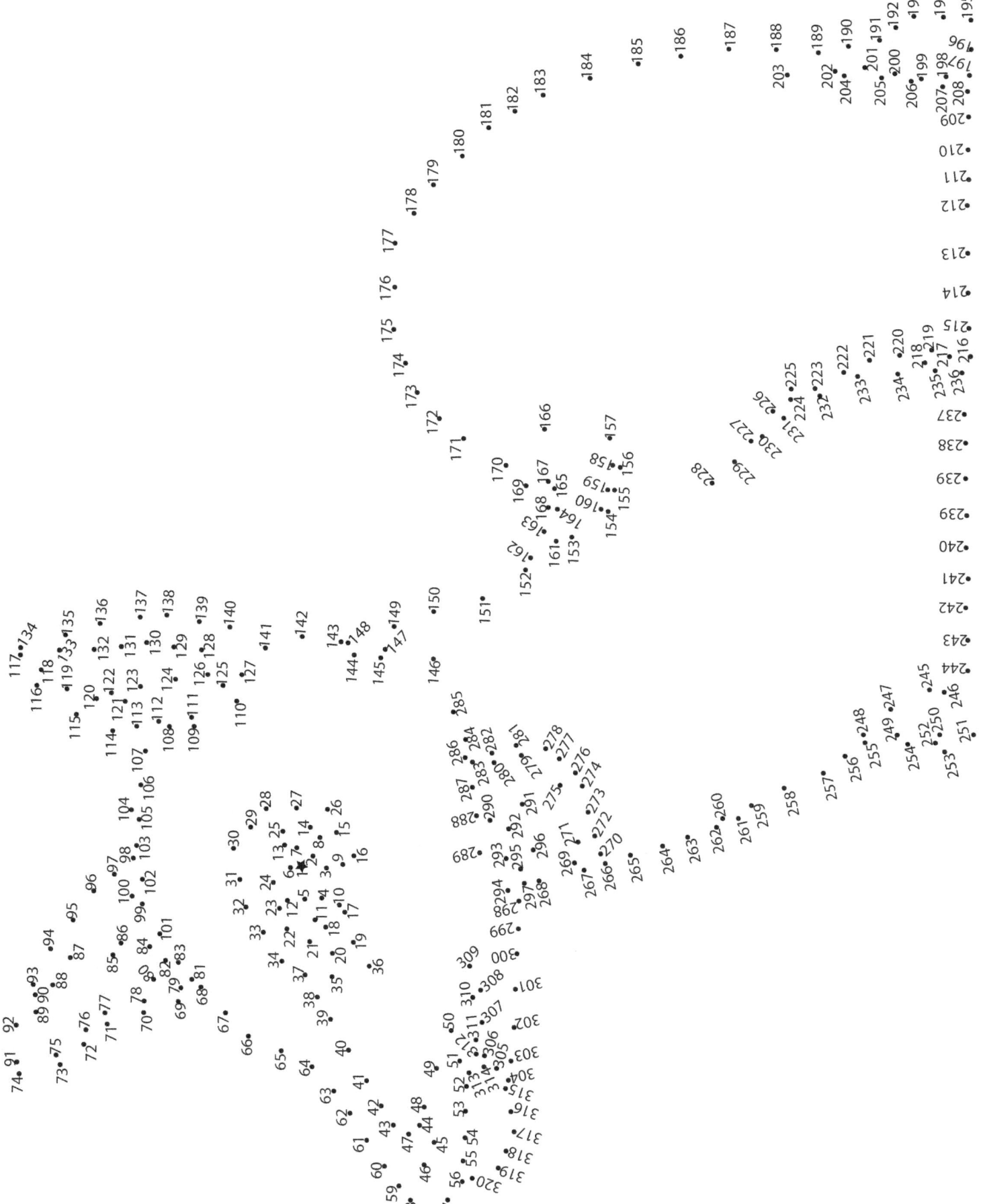

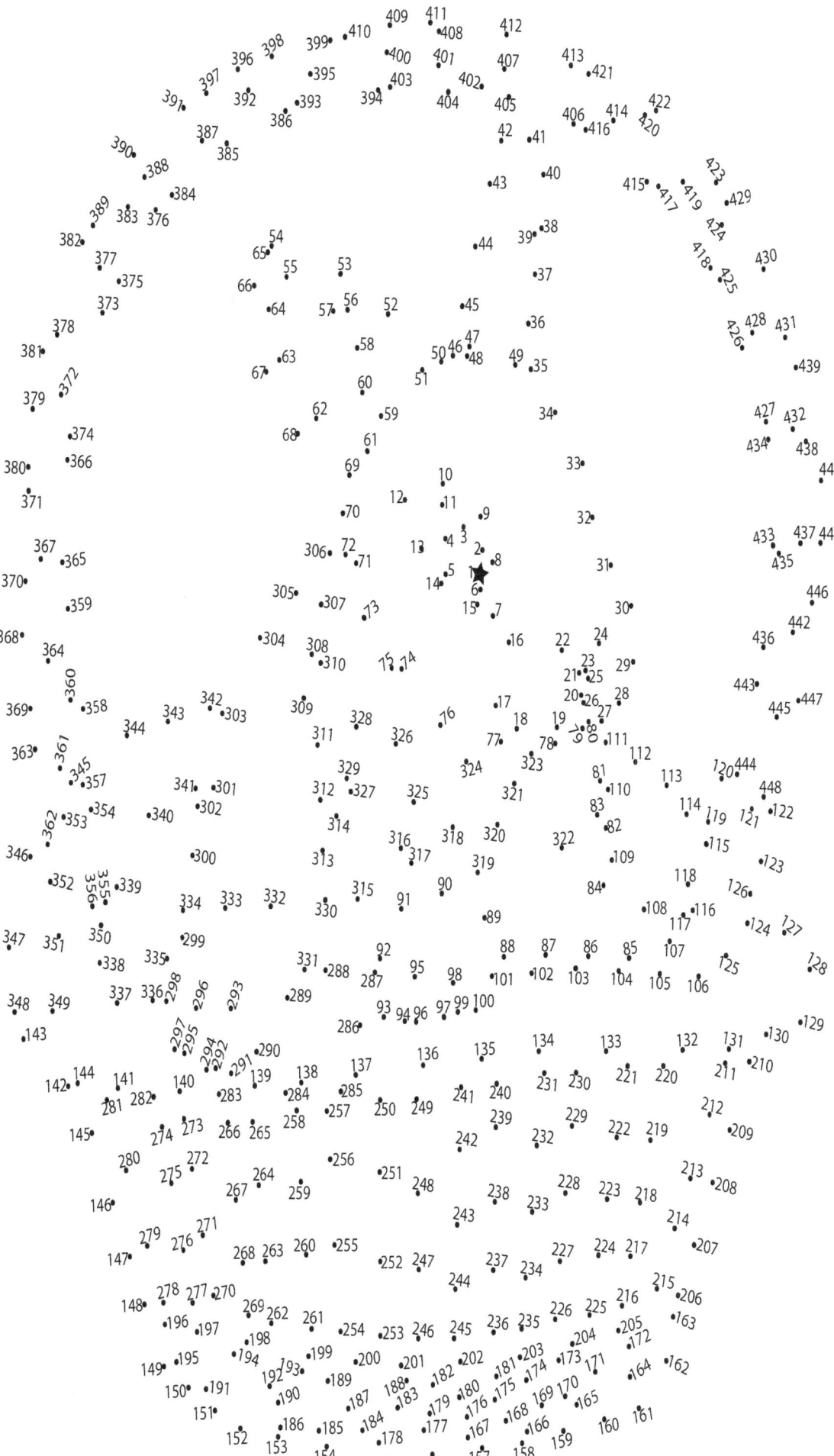

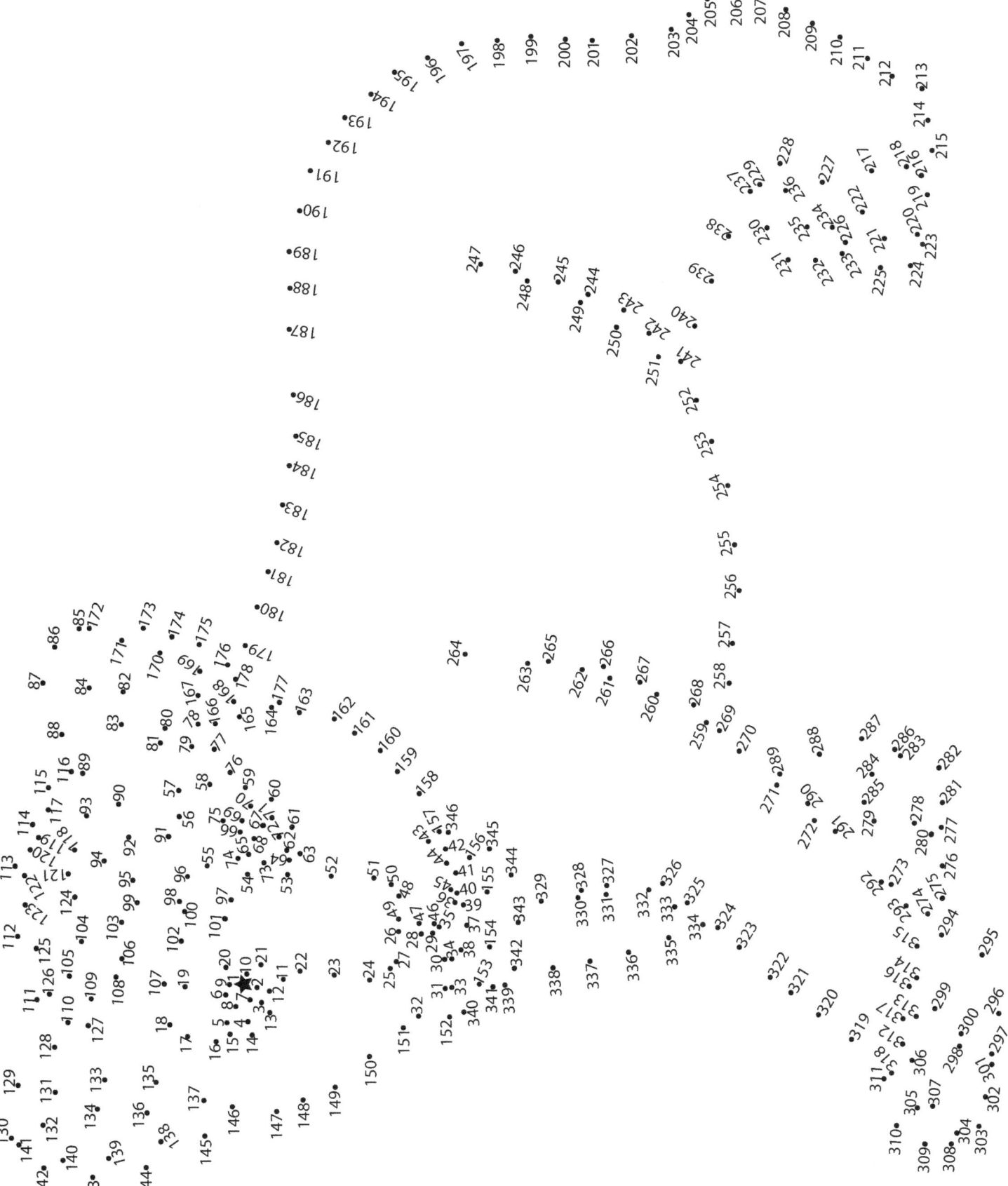

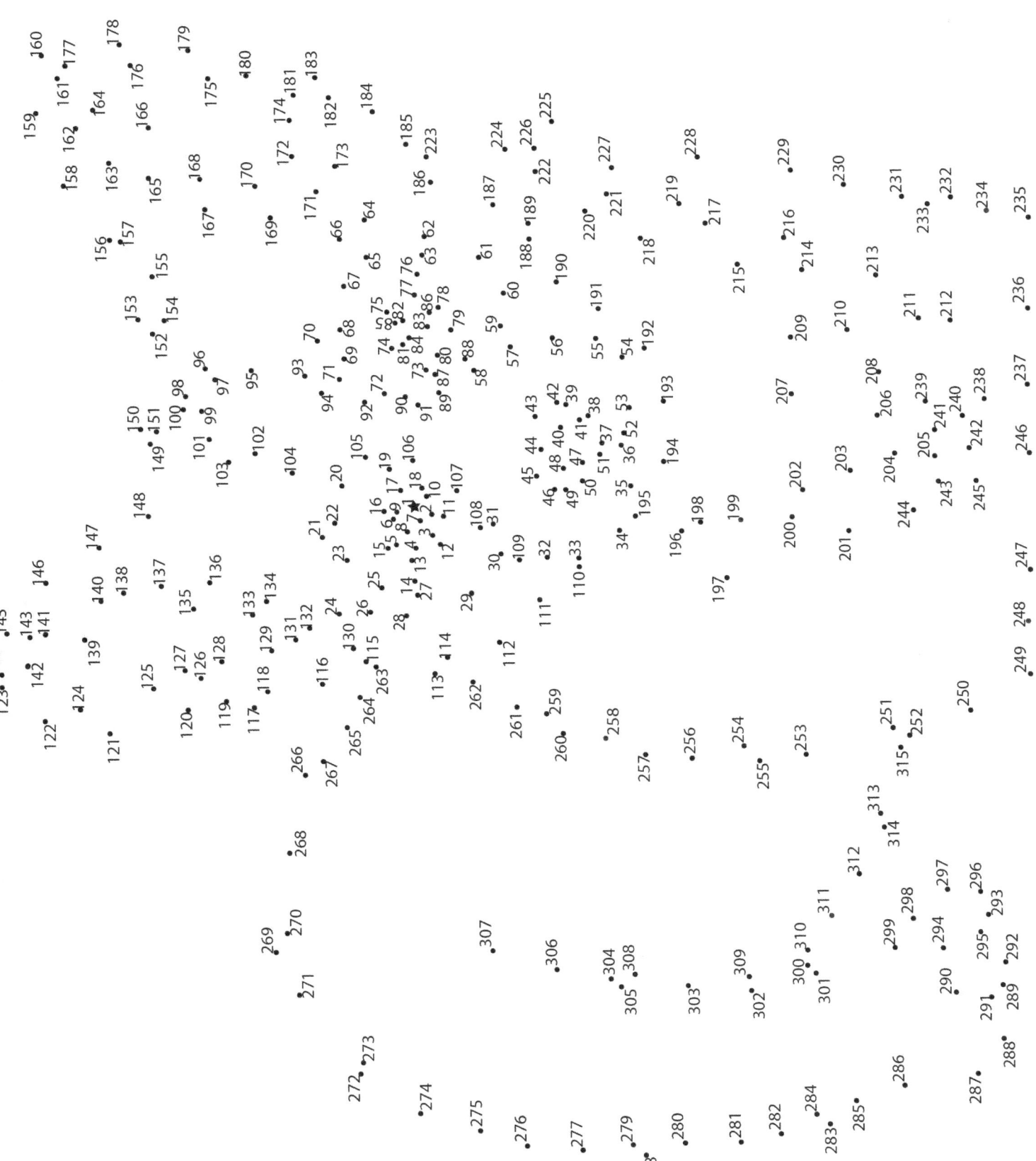

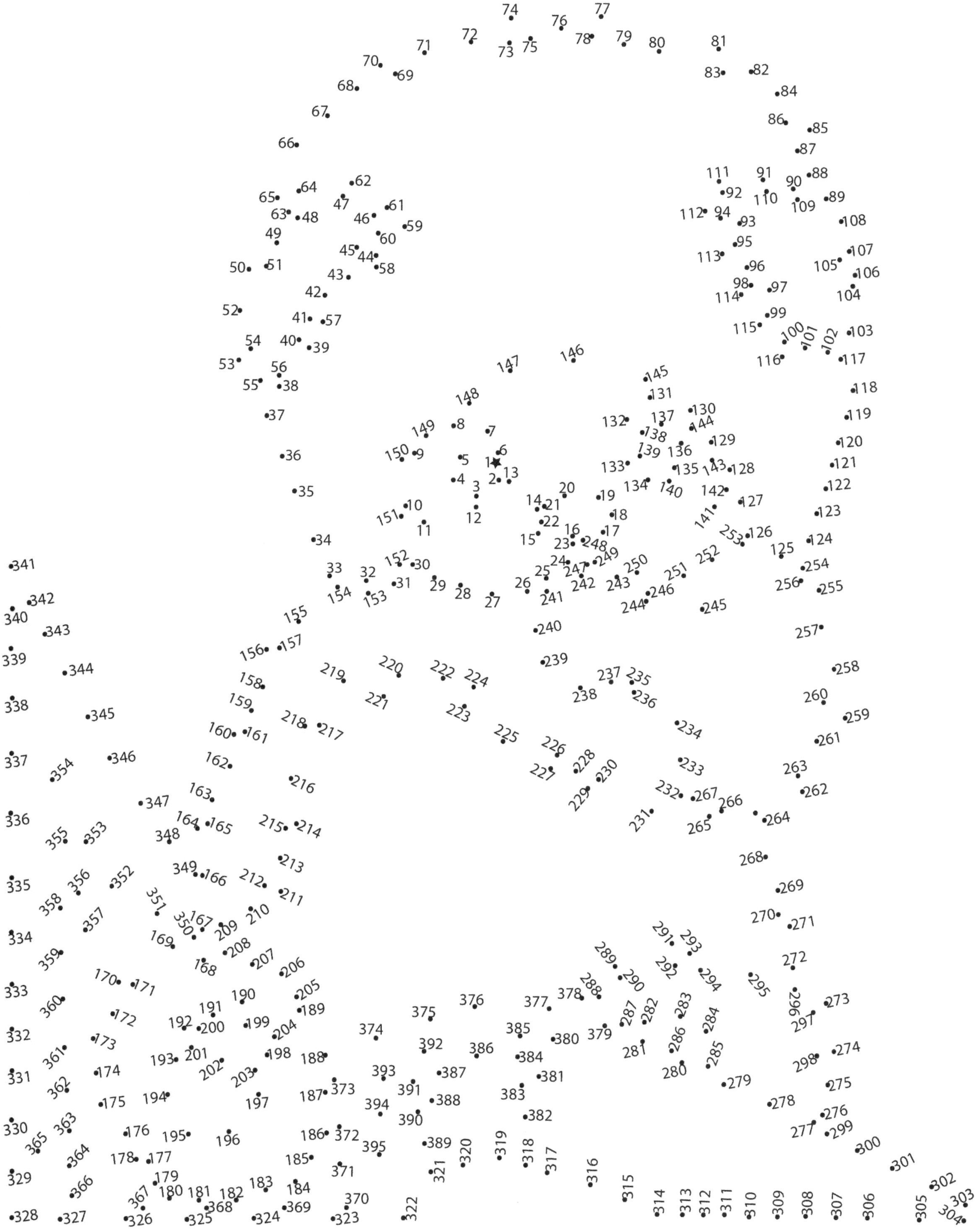

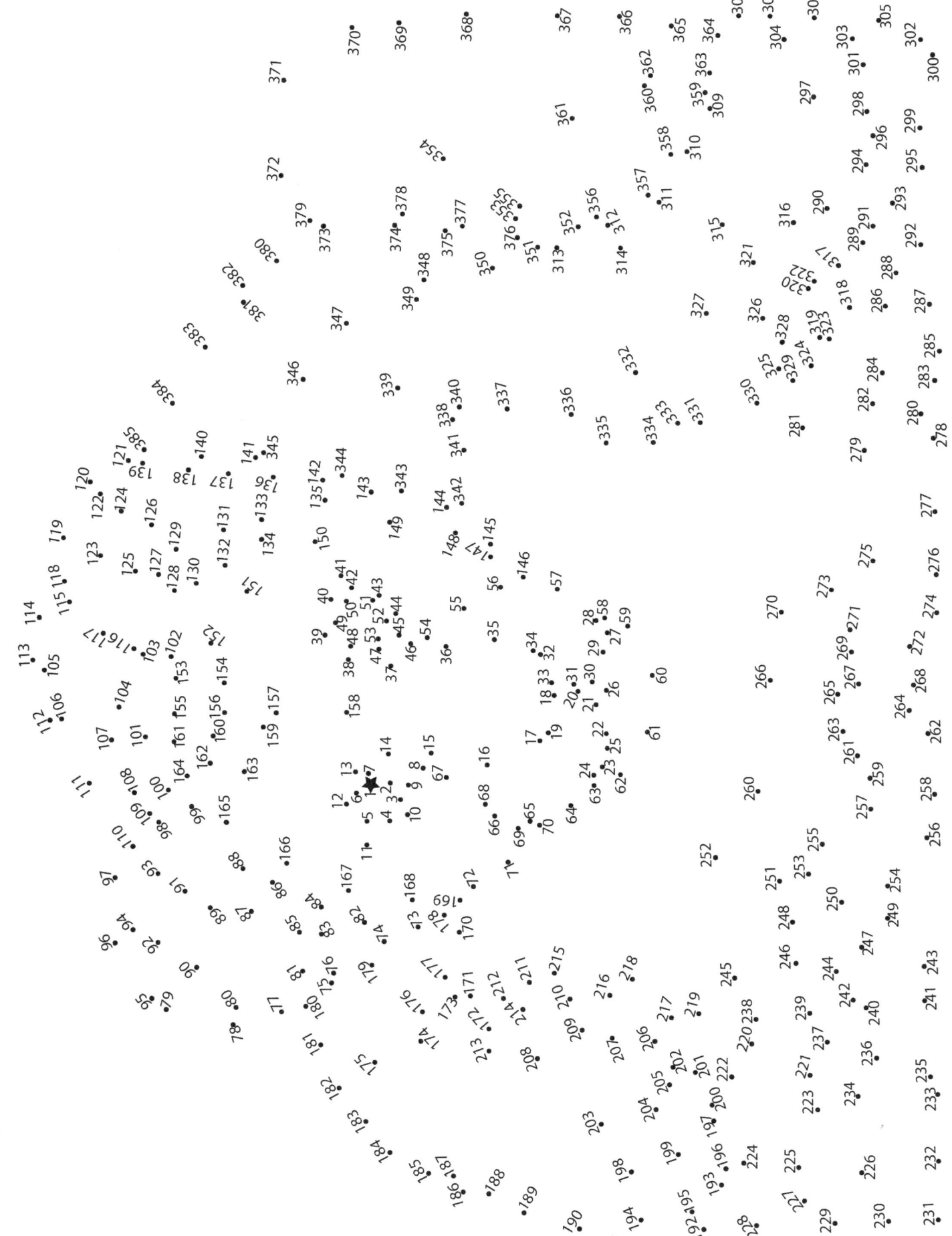

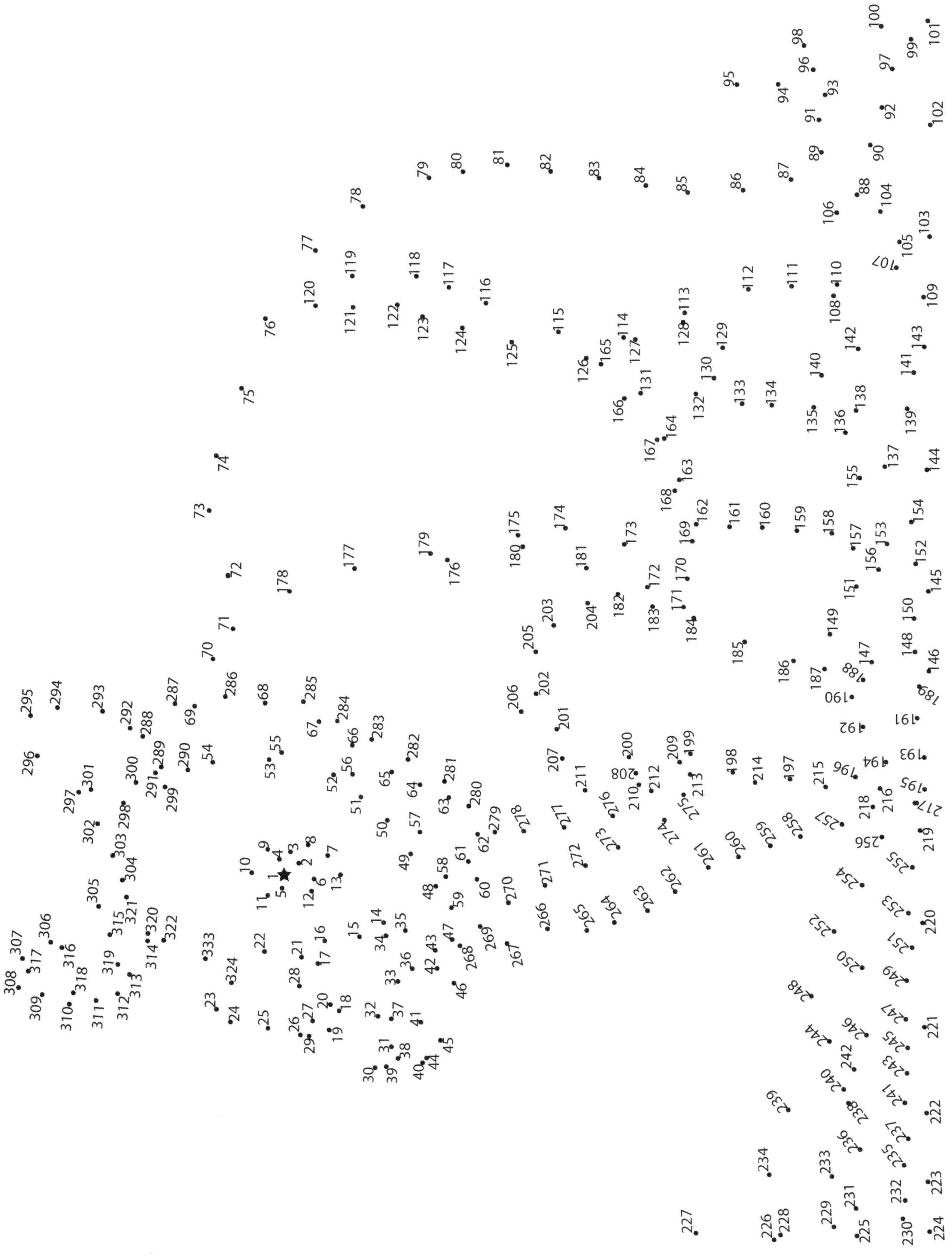

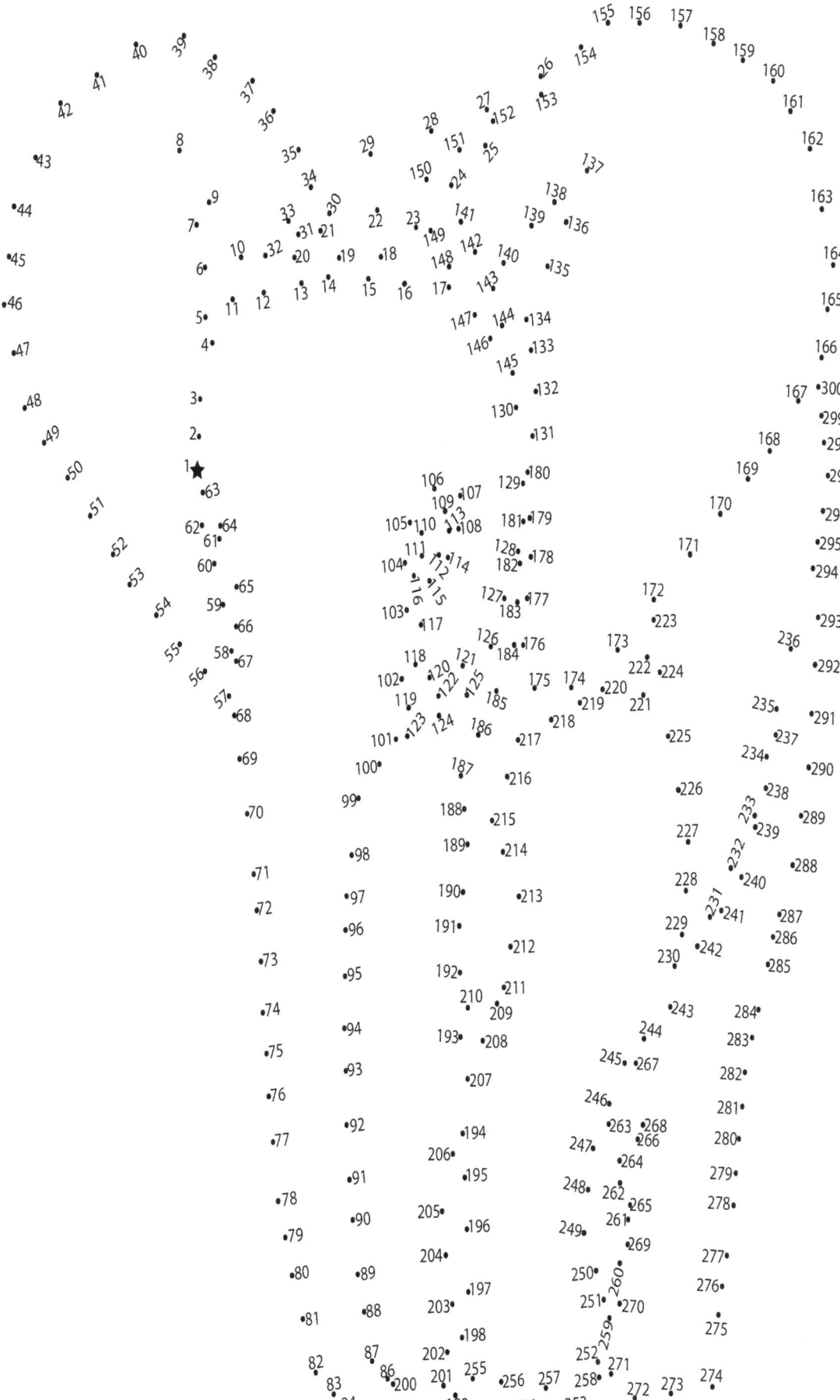

Enjoy bonus images from some of our other fun dot-to-dot books

Find all of our books on Amazon

Beautiful Flowers and Butterflies
Dot-to-Dot for Adults

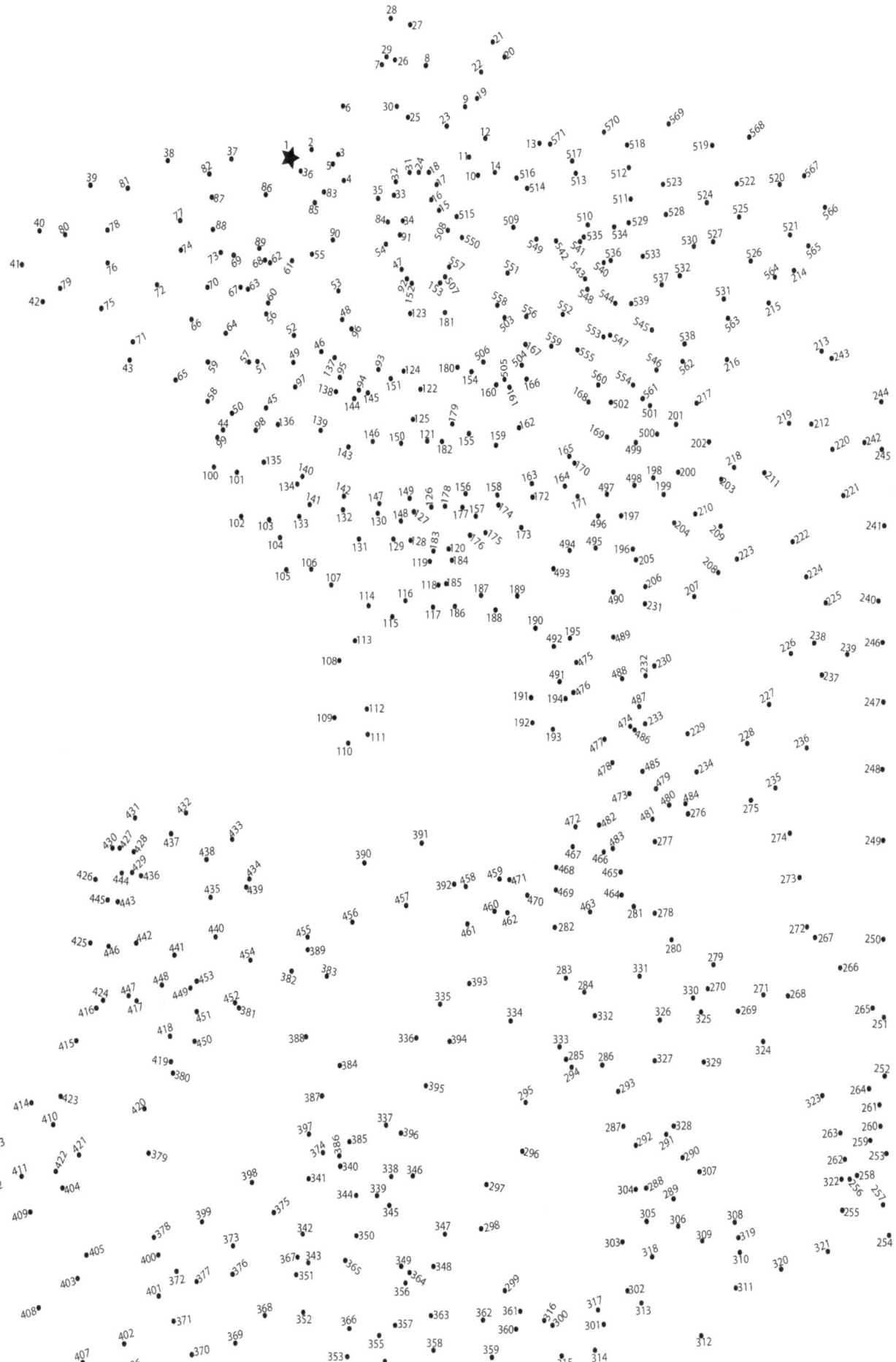

Famous Places Around the World
Dot-to-Dot Book for Adults

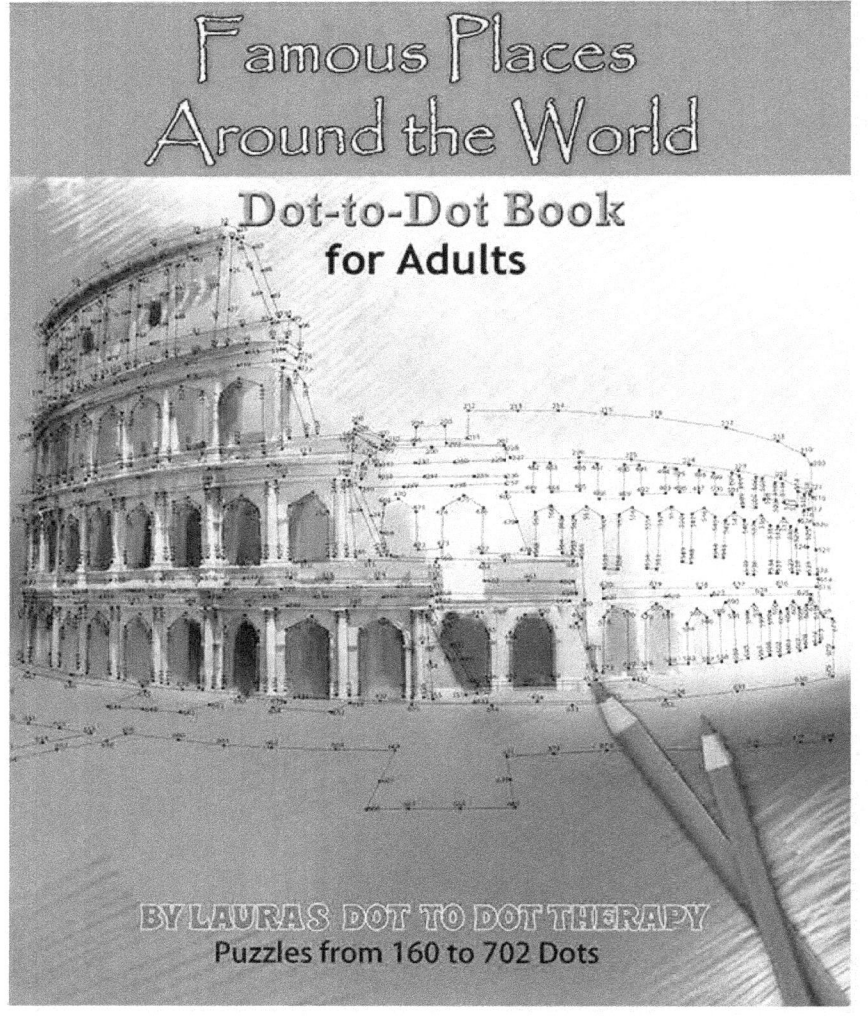

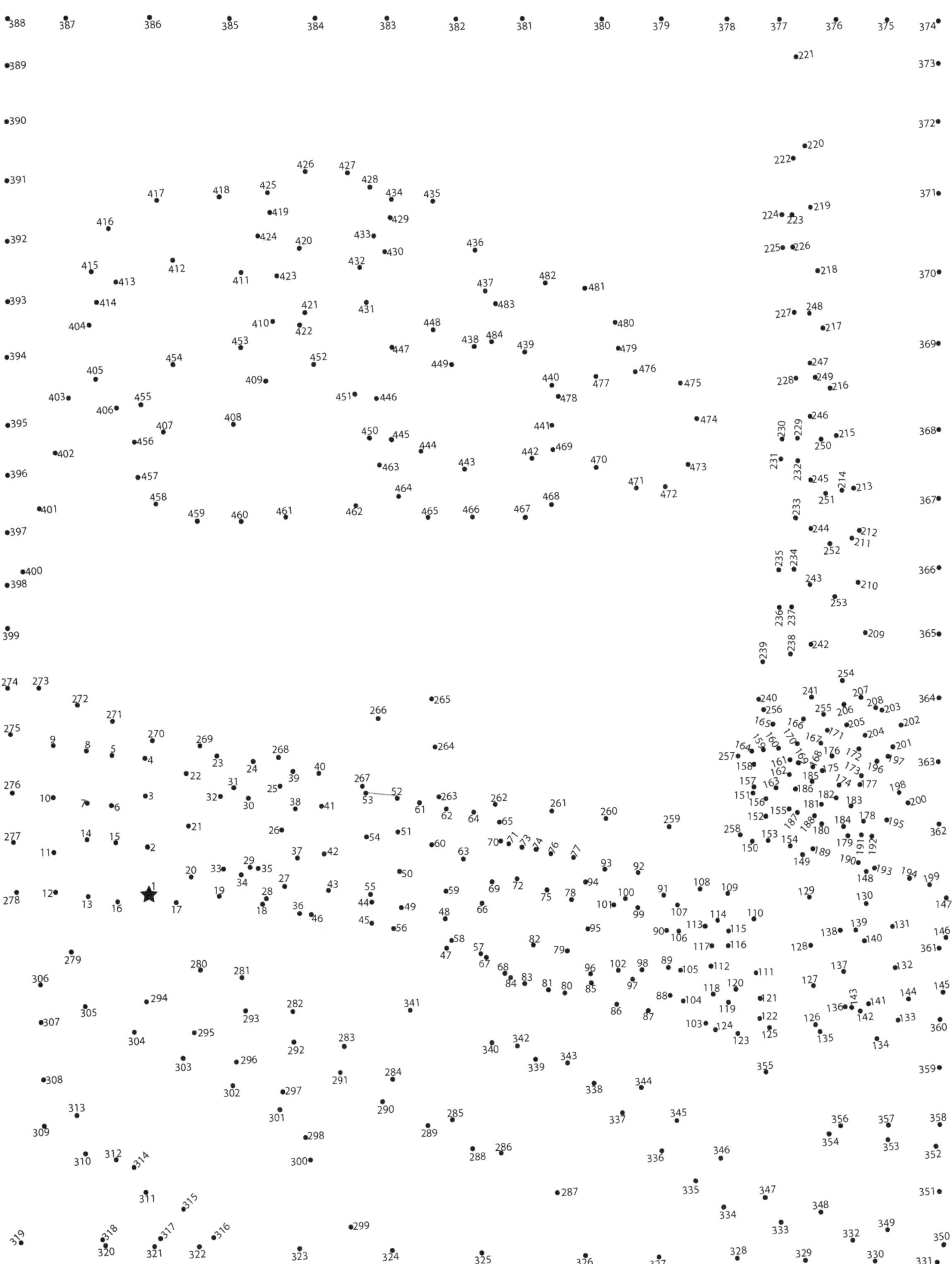

Follow along with the
page numbers from top left
to bottom right

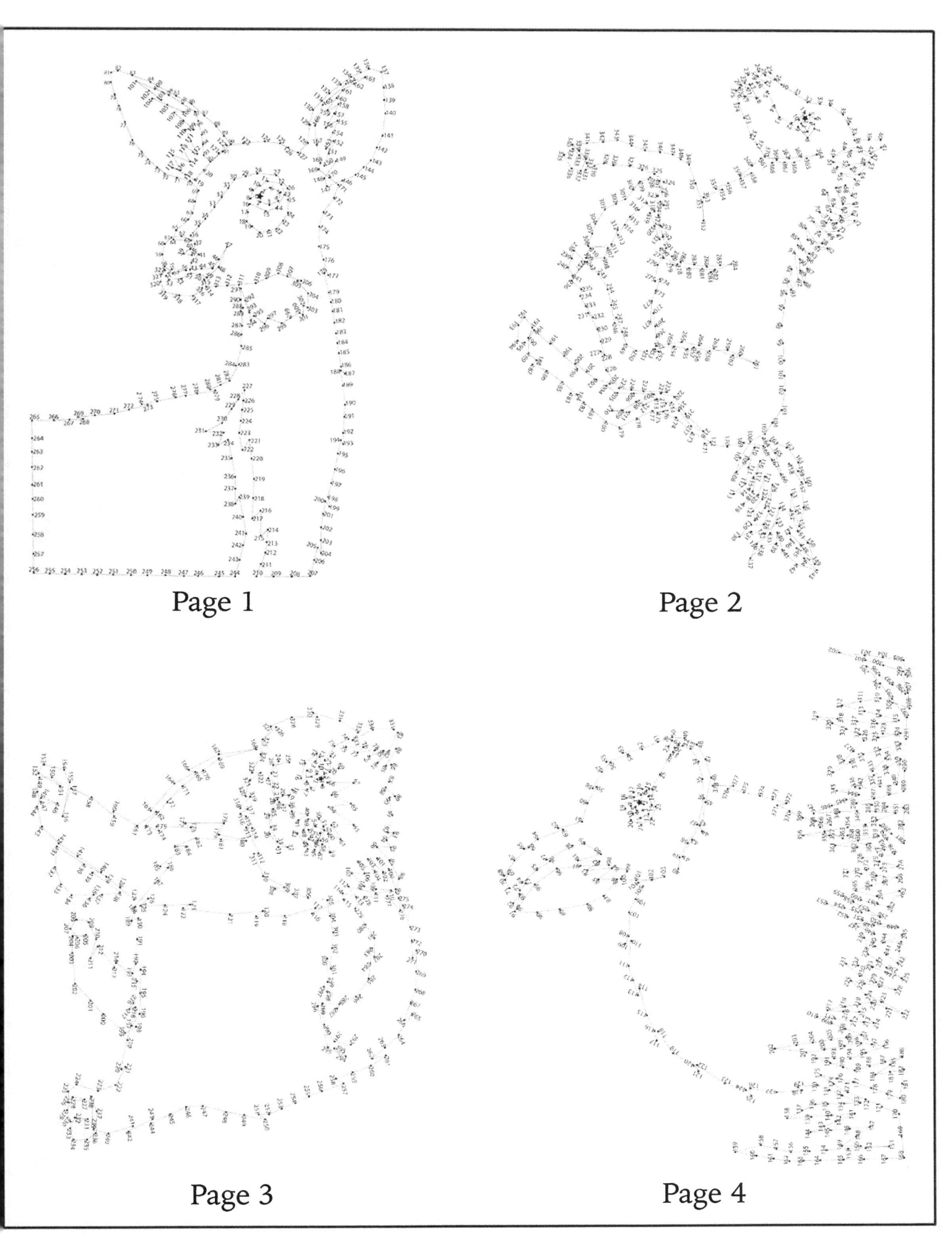

Page 1

Page 2

Page 3

Page 4

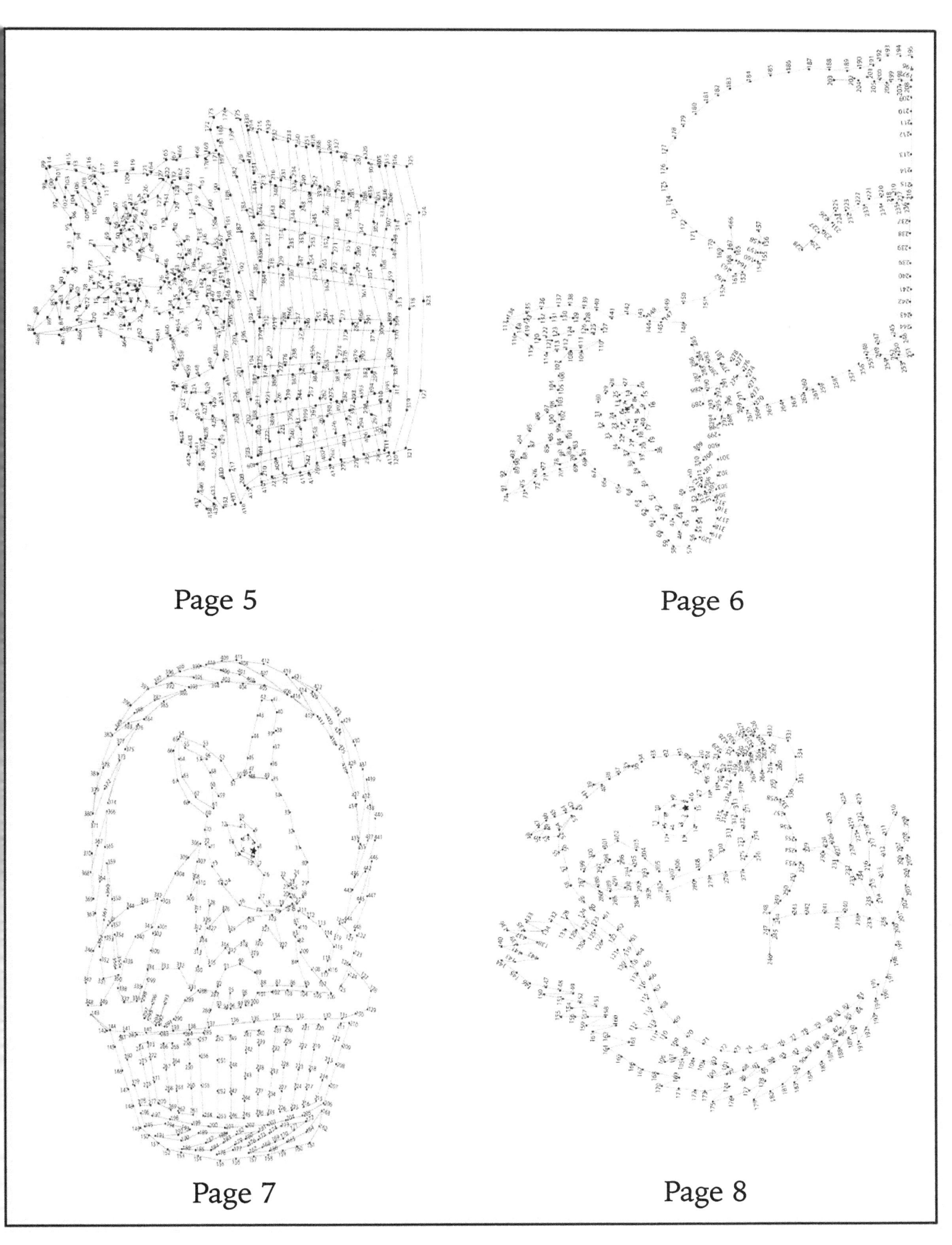

Page 5

Page 6

Page 7

Page 8

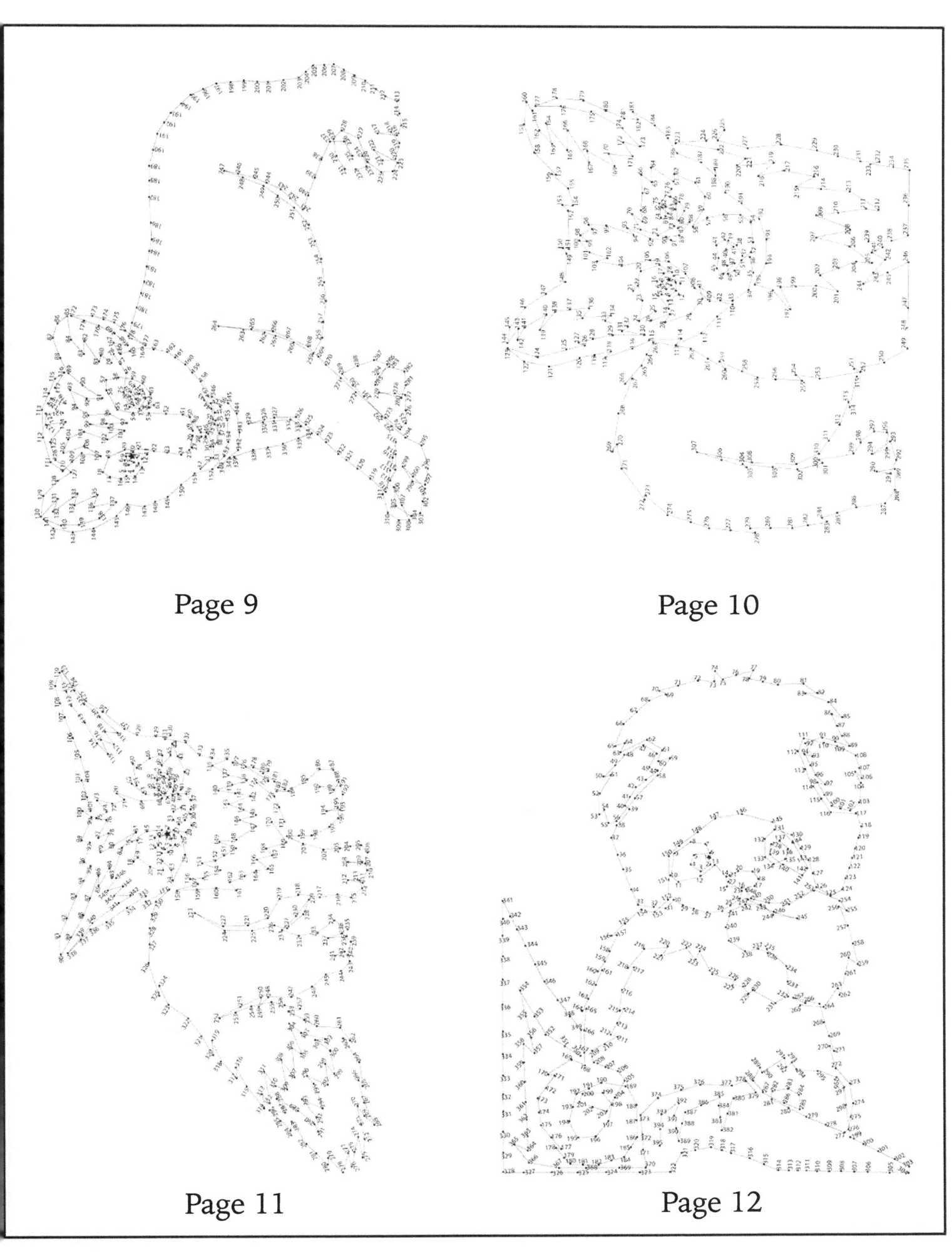

Page 9

Page 10

Page 11

Page 12

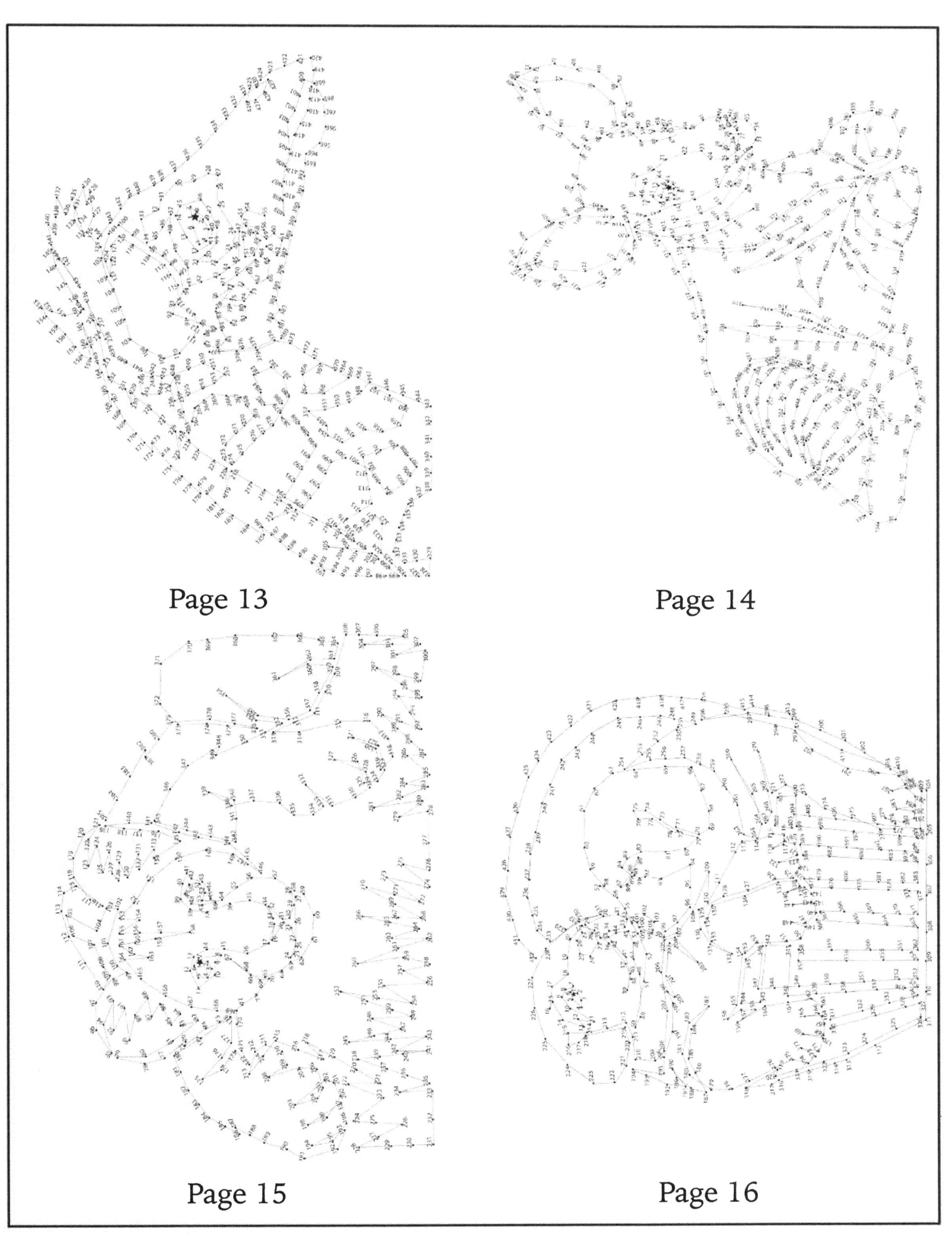

Page 13

Page 14

Page 15

Page 16

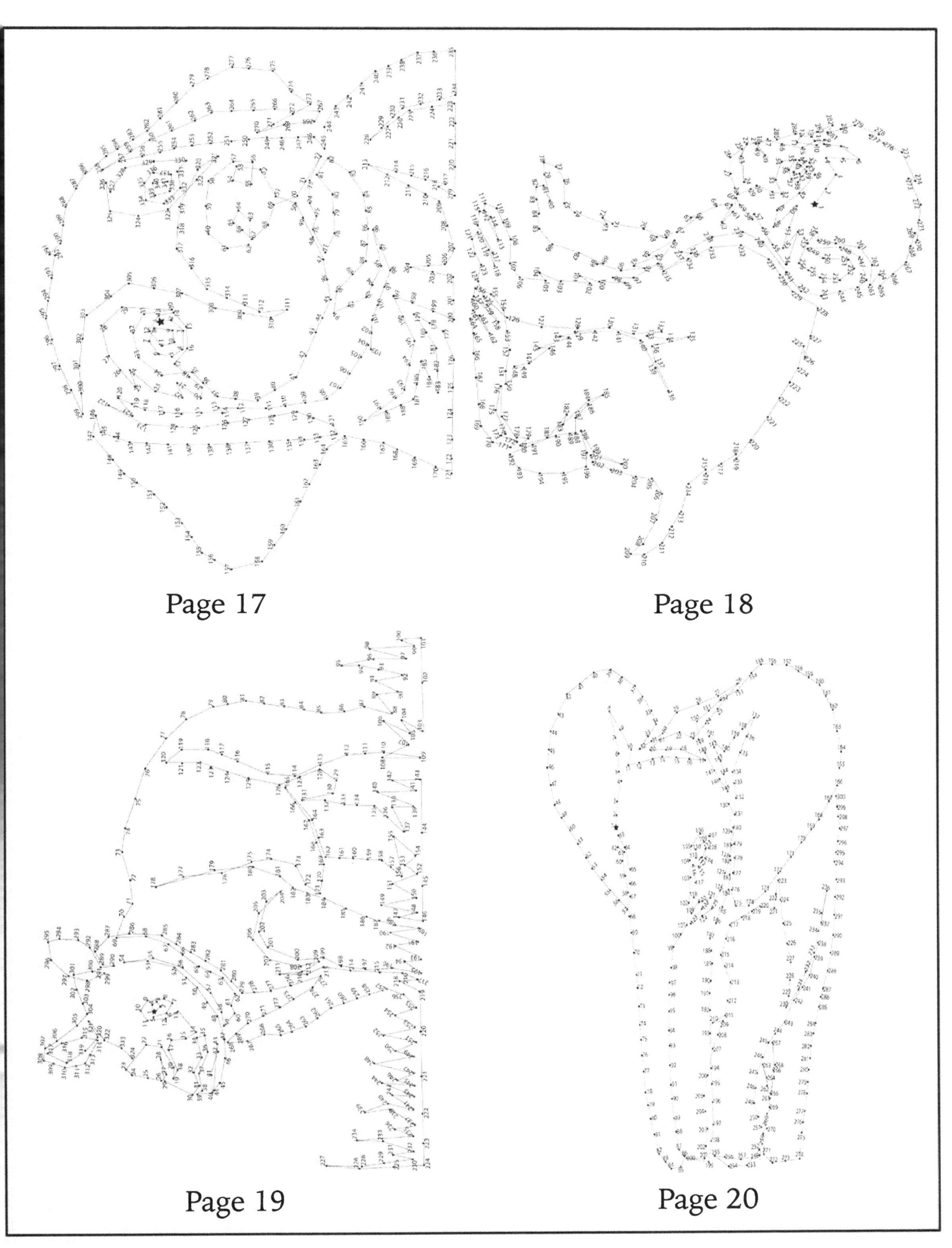

Page 17

Page 18

Page 19

Page 20

www.ingramcontent.com/pod-product-compliance
Lightning Source LLC
Chambersburg PA
CBHW081617220526

45468CB00010B/2919